PRESERVED PRE-1948 RAILWAY CARRIAGES

Royston Morris

I0892642

AMBERLEY

First published 2020

Amberley Publishing
The Hill, Stroud,
Gloucestershire, GL5 4EP

www.amberley-books.com

Copyright © Royston Morris, 2020

The right of Royston Morris to be identified as the Author of this work has been asserted in accordance with the Copyrights, Designs and Patents Act 1988.

All rights reserved. No part of this book may be reprinted or reproduced or utilised in any form or by any electronic, mechanical or other means, now known or hereafter invented, including photocopying and recording, or in any information storage or retrieval system, without the permission in writing from the Publishers.

ISBN: 978 1 4456 9718 5 (print)
ISBN: 978 1 4456 9719 2 (ebook)

British Library Cataloguing in Publication Data.
A catalogue record for this book is available from the British Library.

Typeset in 9pt on 12pt Celeste.
Typesetting by Aura Technology and Software Services, India.
Printed in the UK.

Contents

Introduction

Preserved or heritage railways, as they are now known, have, in the majority of cases, at least one working steam locomotive and either a diesel unit or a diesel locomotive. However, the main thing these railways need to keep them going is the general public coming along and riding the trains.

It's all very well having a decent size fleet of working locomotives, but without any carriages to carry people in, you might as well just call your railway a museum. Therefore, the carriages on a heritage railway are just as important as the locomotives.

One of the biggest attractions for the public is the opportunity to ride in an old-style carriage in its almost original condition and bring back nostalgic memories of their childhood in many cases, especially if a steam locomotive is at the front of the train. There are a number of heritage railways that have restored some of those carriages and operate them on a regular basis on their trains. These carriages have a certain character about them, unlike the modern carriages used on the railways today.

Some of these carriages are bodies that have been rescued and mounted onto suitable donor underframes or brand-new purpose-built ones. They are then lovingly and painstakingly restored by an enthusiastic group of people – this can sometimes take several years but the end result proves to be well worth it.

This book serves to document some of those carriages that have been lovingly restored to their former glories to be enjoyed by a wide spectrum of people nationwide.

Chapter 1

Pre-Grouping Carriages

Companies Amalgamated into the Great Western Railway (GWR)

January 1923 saw the initiation of the government's Railways Act of 1922, which was aimed at amalgamating over 120 independent railway companies into four major railway companies throughout England, Scotland and Wales. The Great Western Railway, however, was not a new company (unlike the other three), due to having been originally founded in 1833.

The area covered by the GWR was the western part of the country from Cornwall to Birmingham and the largest part of Wales. At the grouping there were some forty-three smaller companies which fell under the GWR umbrella. Although there are numerous survivors from some of those companies, they are mostly as grounded bodies. There are, however, several surviving carriages from the original Great Western Railway that were built between 1833 and 1923.

One of the oldest surviving carriages built by the original GWR is this 1881 clerestory-roofed family saloon, No. 248. It was used by King Edward VII when he was the Prince of Wales. At the time of this photograph on 24 March 2007, it was at the Brean Flour Mill in Gloucestershire for restoration.

Built at Swindon Works in 1887 was No. 820, a broad-gauge clerestory-roofed tri-composite carriage. It is preserved as part of the National Collection and was seen on 17 April 2006 at Locomotion, the NRM at Shildon.

This four-compartment brake third was built in 1891 and numbered 416. In 1935 it became a camping coach, before entering preservation in 1968. In 1972 it was moved to Didcot Railway Society for full restoration, which was completed in 2010, and was the location where it was pictured on 17 November 2018.

Noted on 25 May 1996 at Buckfastleigh, on the South Devon Railway, was this 1894-built clerestory roofed family saloon, No. 249. Later it was converted to a directors' saloon, and a departmental engineer's saloon. Withdrawn in 1963 and preserved two years later, it has been subsequentially restored back to its original condition. Note the different style of clerestory roof compared to previous ones in this chapter.

In this 25 August 1997 picture, taken at Big Pit sidings on the Pontypool & Blaenavon Railway, is an 1896 Swindon-built clerestory roofed first-class open brake, No. 231. It was later converted into an engineer's saloon. It was acquired for preservation in 1967, moving to its current location in 1993. Years of open storage have taken their toll on the paintwork.

Soaking up the winter sunshine on 17 November 2018 at the Didcot Railway Centre, Oxfordshire, was this 1901-built third-class auto-trailer, No. 92. It was preserved by the Great Western Society in 1969 but remained in Swindon Works until 1972, later languishing in sidings at Taunton until 1977, when it moved to Didcot.

Built at Swindon in 1901, unusually as a wagon lot, was this royal clerestory-roofed dynamometer car, numbered 790. In 1904 all the equipment was removed at Swindon. It entered preservation in 1965 and on 25 May 1996 it was noted at Buckfastleigh, on the South Devon Railway.

No. 1941 was built in 1901 as a clerestory-roofed third-class carriage and it remained in revenue-earning service until 1945, when it was taken into departmental service and was used until purchased by the Great Western Society in 1966. It was pictured at the Didcot Railway Centre on 17 November 2018.

17 April 2010 saw this four-wheeled composite, No. 290, which was built in 1902, as it undergoes restoration work at the Didcot Railway Centre. This vehicle was transferred into the departmental fleet in 1942, where it was to remain for thirty years until its preservation. Note the oak framework, which was favoured by the GWR.

No. 975 was built in 1902 as a five-compartment third-class carriage that was transferred into the departmental fleet in 1939. It was acquired by the Great Western Society in 1968 and promptly taken to their site at Didcot for display use purposes. I took this picture of it there on 17 April 2010. A full restoration is planned to enable it to be part of the GWS vintage train.

Preserved (but in need of attention) at the Horsehay Steam Trust Railway in Telford on 15 July 2017 was this 1906 GWR-built third-class auto-trailer, No. 38. It was preserved in 1972 having been used as an internal user vehicle at Cardiff Canton depot since its withdrawal from service in the 1940s.

This 1911-built first-class saloon was un-numbered and was built by the Swansea Harbour Trust Railway using a redundant underframe from an unknown *c.* 1850s Manchester, Sheffield & Lincolnshire Railway carriage. It was pictured on 29 December 2005 at the Landore store of the Swansea Industrial & Maritime Museum.

Companies Amalgamated into the London, Midland & Scottish Railway (LMS)

The LMS was formed from thirty-four companies. It operated from London through the Midlands, up to Scotland and parts of North Wales. Amongst the companies that became part of the LMS were: the Lancashire & Yorkshire, North London, London & North Western and Midland Railways. The MR, L&YR and LNWR were the largest of those in terms of areas operated and rolling stock.

In 1840 Queen Adelaide (widow of King William IV) was the first member of the Royal Family to travel by train in the UK. This saloon was built by the London & Birmingham Railway in 1842 exclusively for her and was numbered 2. The underframe was built at Euston Works and the body was by coachbuilders Hooper (later associated with Rolls-Royce). Following her death in 1949, it was stored at Wolverton Works until 1963, before it finally ended up at the National Railway Museum in York, where it was pictured on 17 July 2012.

This rather impressive-looking vehicle was built in 1869 by the London & North Western Railway as two separate six-wheeled vehicles for use by Queen Victoria. In 1895 the two bodies were married together and mounted onto a new underframe and renumbered 802. It was pictured at the NRM, York, on 7 April 2012.

Built at the Bow Works, in East London, of the North London Railway during 1872, was this directors' saloon, No. 32. In 1933 it was renumbered in the departmental fleet series. It was pictured on 17 April 2006 at Locomotion, the NRM at Shildon.

No. 11 was built in 1875 by the BC&W Co. as a five-compartment third-class for the small but highly profitable Maryport & Carlisle Railway. It is the only item of rolling stock from the railway that survives. In the 1930s it was acquired by the National Coal Board and was preserved in 1960. It was pictured on 10 September 2016, at the Chasewater Light Railway in Brownhills, Staffordshire, where it continues to be an ongoing slow but gradual restoration project.

Having its final coat of paint applied on 26 October 2013, following its restoration at the Ingrow workshops of the Keighley & Worth Valley Railway, Yorkshire, is this luggage compartment composite. Built at Derby for the Midland Railway in 1886, it was given the number 358. Like many carriages, it was taken into the departmental fleet in the 1950s and became a Signal & Telegraph stores van before it was preserved in 1968.

Built in 1892 at Wolverton Works in Buckinghamshire, as part of the West Coast Joint Stock as a first-class restaurant dining car was No. 41. In 1933 it was converted to an inspection saloon in the departmental fleet where it was rebuilt as an observation saloon in 1960. Withdrawal came in 1971. On 22 September 1991, it was part of the 'Queen of Scots' train and was seen at Brighton station having worked in on a special excursion railtour.

Built in 1899 by the Ashbury Railway Carriage Company for use by the directors of the Manchester Ship Canal Company was this wooden-underframe un-numbered saloon. In 1902 it was transferred to the Bass Brewery in Burton-upon-Trent for transporting VIPs around the site. It was pictured at the onsite Museum of Brewing there on 4 August 2008.

Undergoing restoration at Beamish, on 19 October 2014, was this 1909-built Highland Railway clerestory saloon, No. 58A. It was built for and used exclusively by the 4th Duke of Sutherland at Dunrobin Castle, before being sold in 1950 to the Romney, Hythe & Dymchurch Railway. In 1965 it was shipped across the Atlantic to Fort Steele Historical Museum in Victoria, British Columbia, before finally being repatriated for restoration at Beamish in 2011.

No. 247 was built in 1910 at Newton Heath as an eight-compartment two-lavatory third-class carriage with corridor for the Lancashire & Yorkshire Railway. In 1917 it was converted to a staff car for a US continental ambulance train before being rebuilt, in 1923, by the LMS as a staff medical examination and eyesight test car. It was later used as an engineers' saloon until preservation came in 1972. On 22 February 2014, it was located in the Manchester Museum of Science & Industry.

Built at Newton Heath, Manchester, in 1911 for the L&YR, this vehicle was completed as a first-class family saloon numbered 50. During 1952 it was gutted and fitted with a generator in the departmental fleet. It was preserved in 1970 and was seen on display inside the Matthew Kirtley Museum, at the Midland Railway Centre, on 16 September 2017.

This Wolverton-built 1920 LNWR corridor composite carriage, No. 2997, was noted on 19 July 2014 at the Railway Age Heritage Centre in Crewe, Cheshire. It was preserved in 1978 and there were plans to restore it as a proposed museum, but this never materialised. It was offered for disposal from Crewe in 2011, but it was still there three years later.

Built at Derby Works in 1921, this was Midland Railway non-corridor elliptical-roofed third-class brake No. 1250. It was withdrawn from revenue service in 1958 and converted to a mobile test laboratory. On 31 August 1998, it was seen at Butterley on the Midland Railway Centre. Note the number of ropes used to secure the tarpaulin to the vehicle.

Companies Amalgamated into the London & North Eastern Railway (LNER)

The LNER was the second largest (behind the LMS) of the four newly formed companies, being made up from eleven major constituent companies. It operated from London (including East Anglia) all the way up to Scotland on the eastern side of the country. The larger companies that came under the LNER umbrella included: the North Eastern, Great Central, Great Northern and North British Railways.

The Eastern Counties Railway was formed in 1836, operating between London and Yarmouth. Pictured on display on 7 April 2012, at the NRM, York, was this 1852-built three-compartment first-class example, No. 64. In 1880 it was rebuilt as a permanent way engineers' saloon and remained in use as such until 1898. It was preserved by the GER in 1899, before being used for displays at various locations in the 1930s by the LNER.

The Manchester, Sheffield & Lincolnshire Railway was formed in 1847. Built at Gorton, in Manchester, during 1876, this was luggage compartment tri-composite No. 154; it was gutted during the 1940s and used as a stores van. It was preserved in 1961 and was seen on 26 October 2013, following an intensive restoration at the Ingrow workshops of the K&WVR.

Looking resplendent on 15 August 2010, at the Buckinghamshire Railway Society in Quainton Road, is newly restored four-compartment brake third-class carriage No. 1470. It was built for the Great Northern Railway, at Doncaster Works, in 1889 and in 1938 was transferred to the departmental fleet. It was withdrawn and privately preserved in 1963, prior to moving to its current home in 1969.

The Lancashire, Derbyshire & East Coast Railway was opened in 1896, primarily to carry coal from Chesterfield to Lincoln. This 1896 Ashbury-built five-compartment third-class carriage, No. 6, is the only surviving piece of rolling stock from the railway. It was withdrawn in 1938, when it was used as a tool van. It became preserved in 1973 but was stored in Derby Works for two years before moving to its current home at the Midland Railway Centre, Butterley, where it was pictured on 16 September 2017.

Built at Ashbury for the MS&LR in 1898, as a three-compartment composite brake, was No. 1470. It was withdrawn from the LNER in 1943 and sold to the shortest standard gauge railway in Britain, the 2-mile-long Easingwold Light Railway in Yorkshire, where it was in service until 1956. It was seen undergoing restoration on 17 November 2012 at the Chasewater Light Railway.

Soaking up the summer sunshine on 17 July 2006 at the Living Museum of the North, in Beamish, was this 1903 North Eastern Railway-built seven-compartment clerestory-roofed luggage composite, No. 3071. In 1949, following the nationalisation of Britain's railways, it was sold to the National Coal Board, prior to being purchased for preservation in 1968.

The BRC&W Company built this eight-compartment suburban third-class vehicle for the Great Central Railway in 1905, numbering it 793. It was to remain in service until 1955, when it was transferred to the internal user fleet and allocated to Hull Docks – it remained there until preserved by the Great Central Railway PLC in 1988. The interior was gutted, and the bodywork was cosmetically restored in 1991, then it was deemed as surplus and has been stored at Swithland sidings, where it was pictured on 21 November 2009.

This non-corridor lavatory third-class carriage is believed to be the last known North British Railway vehicle of its type that remains intact. Built in 1905, at Cowlairs Works in Scotland, and numbered 1748, it was later used as a mess and dormitory van. Preserved in 1973, it was deemed too expensive to convert it back to its original condition, so became used as a store on the Strathspey Railway, where it was pictured in the Boat of Garten yard on 2 August 2004.

Built at Darlington, in March 1906, for the NER, No. 3591 was a clerestory-roofed dynamometer car. It was behind *Mallard* in the record-breaking 126 mph run in July 1938. It was honourably preserved by the British Transport Commission in 1951. It currently stands in the Great Hall at the NRM, where it was pictured on 7 April 2012.

In 1885 the Hull, Barnsley & West Riding Junction Railway & Dock Company was opened. It was renamed in 1905 to the Hull & Barnsley Railway. This lavatory brake third-class carriage was built for the railway in 1908 by the BC&W Co. and given No. 5. In 1956 it was transferred to the departmental fleet, before being preserved in 1968. It can currently be seen at Goathland (Aidensfield in the drama series *Heartbeat*) on the North Yorkshire Moors Railway, where it was pictured on a wet and rainy 8 April 2012.

In 1860, representatives of three large companies (GNR, NER and NBR) arranged for new carriages to be built to run between London and Edinburgh under the name of East Coast Joint Stock. Built at Doncaster in 1908 for conveying King Edward VII, this was first-class saloon No. 395. After his death in 1910, it was used by King George V until 1925, when it was converted to a day saloon for use by Queen Mary. It can currently be seen at the NRM in York, where it was pictured on 17 July 2012.

Seen on 18 August 1990, pictured at the Railway Age Heritage Centre in Crewe, this was 1920 Stratford-built Great Eastern Railway directors'/general managers' saloon No. 1. In 1948 it was renumbered in the departmental fleet before being finally withdrawn in 1971. In 1997 it was purchased by the late Sir William McAlpine and taken to his private Fawley Hill Railway in Buckinghamshire, where it resides to this day.

Companies Amalgamated into the Southern Railway (SR)

The SR operated across the south of Britain between Kent and Cornwall and was the smallest of the newly formed companies, being formed from ten independent constituent companies. The companies that came under SR ownership included; the London, Brighton & South Coast, London & South Western, London, Chatham & Dover and the South Eastern & Chatham Railways.

This example was built in 1834, for the London & Southampton Railway (later LSWR), as a four-wheeled tri-composite, and transferred into the ownership of the Bodmin & Wadebridge Railway in 1855. It was out of use from 1886, prior to being preserved by the LSWR in 1895, and plinthed on display at Waterloo station until 1940, when it was stored for twenty years. It was then restored before going to the NRM at York, where it was pictured on 7 April 2012.

This vehicle was also built in 1834 for the L&SR as a two-compartment second-class carriage, and was, like the previous picture, seen on 7 April 2012. It was also out of use from 1886 to 1895, when it was then put on display by the LSWR at Kingston upon Thames until the 1930s, when it was taken to the NRM. Note the wooden 'dumb buffers'.

The London, Chatham & Dover Railway Longhedge locomotive and carriage works at Battersea, London, is where this four-compartment first-class carriage was built in 1880 and given the number 9. It was sold to Woolwich Arsenal in 1915, but it was out of use from 1924 until it was preserved in 1962 and put into store until 1969. It was the first item to be preserved by the Buckinghamshire Railway Centre at their Quainton Road base and was seen there working on a passenger service on 15 August 2010.

This open first-class carriage, No. 177, was built at Ashford Works, in Kent, in 1900 for the South Eastern & Chatham Railway. It was rebuilt in 1908 as an invalid/family saloon, and in 1936 it was sold to the Army for use in the LMR officers' train, prior to its preservation in 1971. It was pictured at Tenterden, following an extensive restoration by the Kent & East Sussex Railway, on 28 July 2018.

The LSWR built this clerestory-roofed composite in 1907 at Eastleigh as No. 70. In 1931 it was rebuilt, and the clerestory roof was removed. During 1943 it was converted to an ambulance coach at the Royal Victoria Military Hospital in Netley. It was then transferred to the Longmoor Military Railway in 1955. It was moved to the Military Port & Training Centre at Marchwood, Hampshire, in 1967, and entered preservation in 1978. It was seen at the Pontypool & Blaenavon Railway Society, on 18 May 2007, where it was awaiting restoration work.

This clerestory restaurant composite was built in 1908 by the LSWR and numbered 76. It was rebuilt as a 'nondescript' saloon in 1931 and the clerestory roof was removed. In 1943 it was converted to an ambulance coach at the Royal Victoria Military Hospital in Netley. It was transferred, in 1954, to the Longmoor Military Railway, where, in 1961, it was converted to a breakdown coach (with a roller shutter door fitted on one end – this was removed by 2004 during restoration work). It was then transferred to Bicester in 1969, prior to being preserved in 1977. It was seen on 25 August 1997 at the Pontypool & Blaenavon Railway Society in South Wales.

This SE&CR 'birdcage' composite brake, No. 1084, was built in 1910 at Ashford Works in Kent. The Southern Railway installed a lavatory and renumbered it in 1928. Withdrawal from service came in 1954, when it was used as a gauging template van. It was preserved in 1978 at the Kent & East Sussex Railway in Tenterden. It was noted on 9 October 2016, at Sheffield Park, on the Bluebell Railway.

This ‘birdcage’ brake composite carriage was built at Ashford by the SE&CR in 1910 as No. 1100. It remained in service with the SR until 1943, when it was sold to the Army for use at the Longmoor Military Railway. It was finally preserved at the Kent & East Sussex Railway in 1970 and was seen at their headquarters, at Tenterden, on 28 July 2018.

No. 1122 was built at Ashford, in 1911, for the SE&CR, as a seven-compartment ‘birdcage’ lavatory brake composite. Following nationalisation in 1948, the birdcage was removed, and it was transferred to the Isle of Wight and remained in service until 1966. It was noted on 29 June 2019 inside the Train Story Discovery Centre on the island’s preserved steam railway.

No. 641 was built at Lancing Works in 1916, for the London, Brighton & South Coast Railway, as a brake third-class carriage. By 1935 the brake area was converted into three compartments and it was transferred to the Isle of Wight by the SR as a third-class carriage the following year. Today it can be seen hauling passengers on the Isle of Wight Steam Railway at Havenstreet, where it was seen ready to depart on a service on 29 June 2019.

In 1838 the London & Southampton Railway began operating and two years later it was renamed to the London & South Western Railway. No. 1282 was built at Eastleigh in 1921 as a corridor brake third. In 1957 it was taken into the departmental fleet as a breakdown train van. In 1989 (ten years after becoming preserved) it was moved to Mangapps Farm Railway Museum, Burnham-on-Crouch, Essex, where it was being used as a ticket office and shop on 15 June 2013.

No. 6164 was built at Lancing Works, in West Sussex, the year prior to the grouping in 1922, for the LB&SCR as an eight-compartment composite. Transfer to the Isle of Wight took place in 1937 and preservation occurred in 1966, some forty-three years later. It is still seeing regular use carrying passengers on the steam railway where it was observed on 29 June 2019.

Chapter 2

Post-Grouping Carriages

Carriages of the Great Western Railway (GWR)

There are numerous vehicles built by the 'new' Great Western Railway which were built during the period that this book covers and that survive. The Great Western Society Railway Centre at Didcot is home to many of these carriages.

Preserved at the Llangollen Railway in Denbighshire, North Wales, and pictured on 13 October 2007, is No. 163, which was built in 1928 as a third-class auto-trailer coach. These vehicles were used in abundance by the GWR on local push-pull trains. They were designed with a driving cab at one end so that they could be coupled to a steam locomotive and therefore allow the driver to control the train without needing to be in the locomotive. It also eliminated the time-consuming job of running the locomotive around the train.

No. 5131 was built in 1928, at Swindon, as a corridor brake third-class carriage. It was used by the GWR until 1960, when it was transferred and converted to a breakdown train staff coach, before being preserved in 1987 by the Bodmin & Wenford Railway, where I took this picture on 24 July 1995, still in its departmental condition.

This 18 May 2007 picture, taken at the Pontypool & Blaenavon Railway, shows one of the more serious challenges that can face preservation societies when attempting to preserve these vintage carriages. Built at Swindon in 1928, as a five-compartment brake third-class carriage, this was No. 5539. This decrepit vehicle was moved to the Gloucestershire Warwickshire Railway in 2007 and seen again in 2017. It continues to deteriorate and is structurally unsound.

Pictured on 28 February 1996, awaiting restoration at Wallingford on the Cholsey & Wallingford Railway in Oxfordshire, is 1930-built GWR third-class auto-trailer No. 174. This vehicle was preserved in 1961, when it became used as an office for a traffic manager. In 2001 it was sold and moved north to the Llangollen Railway for restoration work. The malting factory complex dominating the skyline behind was demolished in the same year.

The Gloucestershire Warwickshire Railway, at Winchcombe, on 16 October 2010, was the location of this picture of 1930 Swindon-built open first-class brake No. 9004. In 1963 it was converted for use as the North Eastern Region chief civil engineers' inspection saloon. It was used until preservation in 1972. In 1996 it was converted into a club saloon for use on railtours.

The Great Western Society has its headquarters at the Railway Centre in Didcot. This 1932-built open first-class restaurant/kitchen ocean saloon, No. 9118, was preserved by the society in 1966 after seeing thirty-four years in active service. In 1935 it was given the name of *Princess Elizabeth*, which it still carries, as seen in this 12 March 1996 picture.

Preservation societies are not always able to restore their stock and it is left to deteriorate. This 1934-built brake third-class carriage, No. 5500, is a prime example, noted at the Pontypool & Blaenavon Railway Society on 18 May 2007. Withdrawn in 1962, it was transferred to Newport Docks as an internal user vehicle. It entered preservation in 1990 in a gutted condition, where it was used as a stores vehicle. Further deterioration continued and in March 2019 it was seen at Wishaw, where it had been reduced to underframe only.

No. 1087, a 1938 Swindon-built corridor third-class, entered preservation in 1969. Prior subsequentially to becoming preserved it spent its final years as part of the Swindon Works test train and was subsequently part of the internal user fleet. On 18 October 1997 it was pictured at Bewdley station, on the Severn Valley Railway, on a passenger train. Note the milk churns on the trolley on the platform.

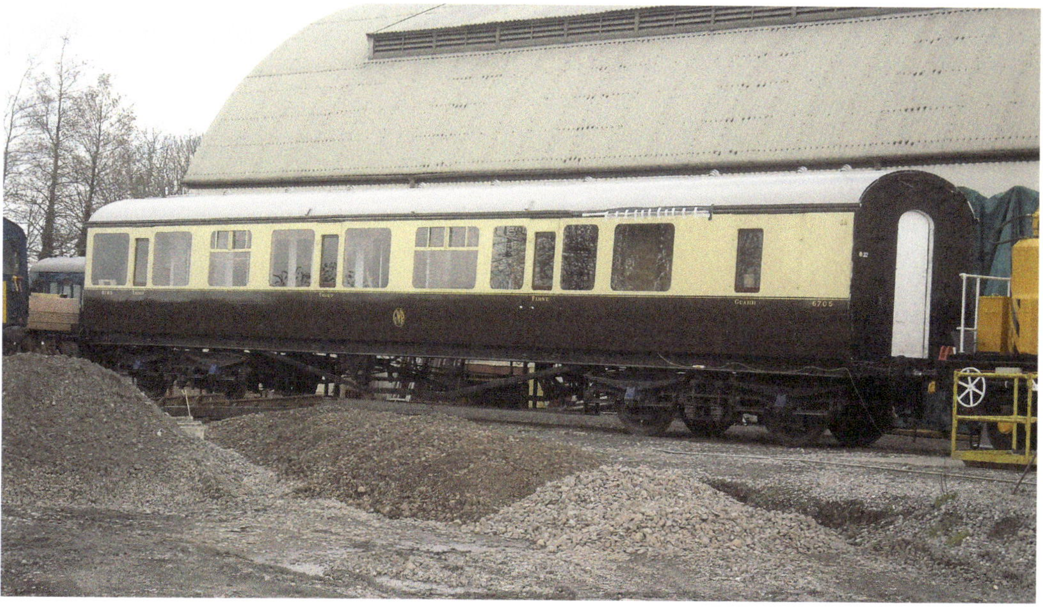

This vehicle, a 1938-built corridor brake composite, No. 6705, lasted in service for twenty-nine years, when it was then withdrawn and sold to the American Steamtown Museum in Vermont. In 2006 it was donated, by the US National Parks Service, to the West Somerset Railway and was shipped from Halifax, Nova Scotia, in March 2007 to the Railway Age Heritage Centre, Crewe. It moved to Williton in 2011, where it was pictured on 5 March 2016 with its restoration virtually completed.

Built at Swindon Works in 1940, this diesel streamlined railcar was numbered 22. These vehicles were introduced in 1933 and a total of thirty-eight were built. Several were withdrawn in the 1950s with quite a few lasting into the 1960s, some being converted as postal/parcels vehicles. This example lasted in service until 1962 and entered preservation the following year. It was pictured at the Great Western Society Railway Centre at Didcot, Oxfordshire, on 17 April 2010.

Third-class corridor No. 796 was built at Swindon in 1947. In 1961 it was converted for use at the Railway Technical Centre in Derby as a dynamometer car. It was purchased for preservation in 1983 and moved to Wolverton Works five years later, where it was restored for main line use. On 12 March 1996, it was seen visiting the GWS Railway Centre in Didcot and was pictured beneath an ominous-looking heavy sky.

Seen soaking up the bright 7 June 2015 sunshine, at Bishops Lydeard station on the West Somerset Railway, is the beautifully restored Swindon-built 1947 Hawksworth Inspection Saloon, No. 80976. The vehicle has only been preserved since 1988 so it is relatively a newcomer in preservation terms.

Carriages of the London, Midland & Scottish Railway (LMS)

At its formation in January 1923, the LMS had inherited some 20,276 carriages. This number had risen to 24,023 by 1934, which included new-build vehicles to replace some of the inherited ones.

This corridor open third-class carriage was built at Derby Works in 1925 as No. 16696. In 1950 it became a staff and mess car in the LMR control train. It was preserved in 1980 and taken to the Swanage Railway in Dorset. By the time this picture was taken on 9 May 1998 it had moved to the ill-fated North Somerset Railway project at Radstock.

The shabby appearance of this vehicle is caused by it having been used as a steam boiler testing van for thirty-six years. It was built by the MC&W Co. for the LMS in 1926 as an open third-class carriage, No. 6039, before being converted in 1951. It was pictured on 8 April 2012, awaiting restoration at Grosmont, on the North Yorkshire Moors Railway. It is unique in that it is the only surviving and working boiler testing van that was converted from passenger stock.

1927 Derby-built open third-class carriage No. 18358 became an Overhead Electrification coach in the 1950s. It was withdrawn in 1984 where it was acquired by the owner of the former Castle Ashby station in Northamptonshire, where it was used as part of the seating of the onsite Dunkley's Restaurant. It was pictured there on 13 October 2012.

Pictured at the secure former MOD storage facility, at Long Marston in Warwickshire, on 10 June 2007 is this 1929 Derby-built open third-class carriage, No. 2924. In 1939 it was converted to an ambulance ward car and the following year it was captured by German forces and used by them for the same purpose. In 1945 the Allied forces recaptured it and it was bought back to the UK and taken to the Longmoor Military Railway. It was finally preserved in 1970, moving from location to location, before finally being acquired by Peak Rail, who plan to restore it as an ambulance car.

On 9 May 1998, the site of the ill-fated North Somerset Railway project, at Radstock, was where the dilapidated-looking shell of 1933 Wolverton-built corridor third-class carriage No. 1535 could be seen languishing. In 1963 it was used as a breakdown train staff and tool coach until its preservation in 1988. It is currently stored and sheeted in Gloucestershire pending a decision regarding its future.

This four-compartment corridor brake third-class carriage, No. 5734, was built at Wolverton Works in 1935. After thirty years of carrying passengers it was used as a civil engineering staff and tool coach. In 1987 it was preserved and was seen (still bearing its departmental number) at Swanwick Junction, at the Midland Railway Centre, on 10 April 1994.

The heavy modifications carried out in 1960, when this vehicle was transferred for use as a tool van in the Buxton-based Breakdown Train Unit, are clearly evident in this 28 April 1996 photograph taken at Darley Dale, on Peak Rail. The railway has kept it in this condition and regularly use it as part of their demonstration works trains. It began life in 1936, at Derby Works, for the LMS, as an open third-class carriage, numbered 9205.

On 16 April 2006, the Marsden Rattler Restaurant in South Shields, on Tyneside, was offering '2 Meals for The Price of One'. This 1937 Wolverton-built nine-compartment third-class carriage, No. 11937, was purchased in 1986. Prior to that it was used as a chief engineer's staff and dormitory coach. It has been completed gutted internally of all its original fixtures and fittings and is used as part of the kitchen area at the restaurant.

During the years prior to the Second World War the railway companies commenced various electrification programmes throughout the country. Among the vehicles introduced at that time by the LMS was this 1937-built second-class open-driving trailer, No. 29289. It was one car from the 650vDC third-rail designated Class 503 units that worked in the Wirral and Mersey areas until 1991. Since its preservation in 1996, it has had the misfortune to have been homed at two museums which have since closed. The last one was the Electric Railway Museum at Baginton, near Coventry, where it was pictured on display on a wet 18 September 2010.

The various companies over the years have always been successful in building luxury carriages for various members of the Royal Family and household staff. Pictured on display in the National Railway Museum, at York, on 7 April 2012 is this 1941 Wolverton-built LMS twelve-wheel armoured Queen's saloon, No. 799.

This LMS district engineers' inspection saloon was built in 1941, at Wolverton, and numbered DM 45046. In 1983 it was withdrawn from service and taken with five others adjacent to the mainline at Dawlish Warren as one of the Western Region Staff Association camping coaches. Here it was fitted with the cast nameplate *Plymouth*. It was pictured on a bright, hot sunny 3 July 2016, when the future of the site was under threat of closure.

No. 2300 was built in Derby, in 1946, as a corridor third-class vehicle and it saw twenty-four years in revenue service. In 1970 it was purchased by Bass Brewery at Tavistock Junction, near Plymouth. It was sold by them as an empty shell, in 1978, to the Severn Valley Railway Association for preservation. Over the next six years it was fully restored and made operational, using the interior from a sister carriage which was scrapped in 1980. It was pictured in passenger service at Bridgnorth on 18 October 1997.

Carriages of the London & North Eastern Railway (LNER)

The LNER fell behind the LMS by a short measure in that they owned 20,000 carriages – although the variety of different types of carriages that they inherited eclipsed their main rival by a ratio of 2:1.

Built just after the grouping, in 1924, at York Works as a corridor composite for use on the 'Flying Scotsman' train services, this was No. 10155. It is a unique vehicle in that it has an extremely unusual half compartment, which contains three seats that all face a blank wall! In 1959 it was rebuilt as a mess and tool van. It was awaiting restoration to its original condition on the Great Central Railway at the time of the photograph. On 21 November 2009 it had been cosmetically restored.

Another 1924 York-built carriage for the 'Flying Scotsman' services was this corridor composite brake carriage, No. 10178. It was pictured inside the Newbridge engineers' shed, on the North Yorkshire Moors Railway, on 8 April 2012. In 1958 it was converted to a staff and tool van and was transferred to Blyth depot as an internal user vehicle in 1979, although preservation followed later that same year.

Pictured on 28 July 1995, at the Bere Ferrers Station Museum, near the Devon and Cornwall border, and still in its 1959 signal and telecommunications department staff and tool van condition, was this 1926 York-built brake third-class carriage, No. 3107. As can be seen, it has had several doors removed as part of the process to restore the vehicle to its original highly polished teak condition.

No. 1211, a first-class sleeper, was the last wooden-bodied sleeper car that saw service in the UK. It was withdrawn in 1972. It was built at Doncaster Works in 1935 and was found at the Boat of Garten yard, on the Strathspey Railway, on 2 August 2004, where the roof was sheeted over to further protect it from the elements.

Looking good at Pickering, on the North Yorkshire Moors Railway, on 8 April 2012, is this first-class restaurant car. It was built at York in 1935 as open third-class carriage No. 43654. In 1961 it was used as an office coach in the ER control train before being preserved in 1980, when it was converted to the guise seen here.

With very faint traces of teak visible underneath the fading pale-blue paintwork, this 1936 York-built open third-class carriage, No. 23981, awaits its turn to be restored to its former glory at Swithland sidings, on the Great Central Railway, in Leicestershire, on 21 November 2009. It entered preservation in 1994 having been purchased from Perth yard, in Scotland, where it had been used as a staff coach.

Pictured at the place where it was built (albeit some sixty-six years later), is this 1937 Coronation Set Beavertail observation saloon, No. 1719. It was in need of some restoration work when pictured at Doncaster Works during a public open day on 26 July 2003.

The teak carriages built by the LNER are a delight to look at, especially when they have been so beautifully restored to their former glory and highly varnished accordingly. This 1937-built buffet car, No. 641, pictured at Pickering on the North Yorkshire Moors Railway on 8 April 2012, clearly shows this. This particular example was preserved in 1974 by the Dart Valley Railway in Devon, before moving to its present pictured location in 1983.

Carrying its 1946 number (and non-authentic GE prefix) of 10078, and pictured in Kidderminster Yard, on 18 October 1997, at the Severn Valley Railway, where it has been resident since its preservation in 1972, is this corridor composite brake, which was originally numbered 24068 and was built at York in 1937. Note the semaphore signal gantry in the background.

This crimson and cream-liveried buffet car, No. 24280, was built at York in 1937 and entered into preservation thirty-six years later. It was pictured between passenger duties on the Great Central Railway, at Rothley, on 21 November 2009.

No. 56856 was built by Metropolitan-Cammell, at Birmingham, in 1938 for the LNER as an open third-class carriage. It remained in service until 1950, when it was used as a medical officers' saloon and remained in that guise until purchased for preservation in 1981. It was pictured following its restoration to its original teak condition at Newbridge, on the North Yorkshire Moors Railway, on 8 April 2012.

This 1939 York-built corridor brake third-class carriage, No. 57451, was preserved in 1965 and in 1970 it featured as the carriage in which the gentleman on the train sat in the hit film *The Railway Children.* It was pictured at Rothley, on 21 November 2009, where it was undergoing external cosmetic restoration for static display. There is not much of the interior that remains intact with no plans to replace it.

Carriages of the Southern Railway (SR)

They might have been the smallest company, but this still saw them own 10,800 passenger carriages in 1923.

Seen located in Ewhurst sidings, on the Kent & East Sussex Railway, on 28 July 2018 was this 1928 Ashford-built corridor composite, No. 5153. It is stored and awaits restoration, which it is in dire need of. It was acquired for preservation in 1965 but fifty years later it remains in a deteriorating condition. The underframe is apparently in a very good condition.

Brake composite No. 6575 was built at Eastleigh in 1929 and was in service until withdrawn in 1960. It was then purchased by the Bluebell Railway and, as such, it was their first acquired preserved carriage. On 9 October 2016 it was pictured at Horsted Keynes, covered with a tarpaulin as it had been withdrawn from service on the railway needing repairs to the roof and brake end. It waits for this work to be carried out.

Built at Eastleigh Works in 1930, this open third-class carriage, No. 1381, remained in service until 1962, when it was converted to a staff and tool van. In 1979 it was preserved by the Swanage Railway, in Dorset, and was pictured there on 15 March 1996.

Noted at Wishaw, Warwickshire, looking in a very poor condition, was this 1930 Eastleigh-built bogie corridor composite, No. 5600. In 1962 it was converted to an inspection saloon before entering preservation with the Kent & East Sussex Railway in 1990. The end, seen in this 13 June 2015 picture, suffered damage from an arson attack in 2007.

This 1931 Southern Railway Eastleigh-built corridor third-class carriage, No. 2356, was withdrawn in 1961 and taken to Streatham for use as a carriage and wagon department internal user mess and stores van. It was acquired by the Bluebell Railway Society in 1972 and taken to Horsted Keynes station, where, on 14 April 1996, it was being used as a bookstore and shop.

Pictured on 28 July 2018 undergoing restoration at the Rother Valley Railway in Robertsbridge was this 1931 Eastleigh-built corridor brake third-class carriage, No. 3690. There is no interior and it will have a brake plus one compartment with the remainder as an open saloon and exhibition space. In 1962 it was used as part of the SR control train until 1980, then it languished away in a siding for eight years before being purchased.

This 1932-built corridor brake third-class carriage, No. 2768, was transferred to the departmental fleet in 1962 until it was bought by the Swanage Railway in 1981. It was seen in Eldon Sidings, just outside of Swanage, on 11 January 2002, where the bodywork was in such a bad condition it was being held to the underframe by ratchet straps.

Open third-class No. 1336 has a chequered history to it. It was built by the SR at Eastleigh in 1933 and operated until 1962. It was then transferred to the internal user fleet as a mobile office. In 1970 it was transferred to the departmental fleet (unusually, as that normally happens the other way around). Preserved in 1988, it was given a complete restoration and it was back in traffic by November 2008. It was seen at Horsted Keynes shed, on the Bluebell Railway, on 9 October 2016.

No. 4432 is an unclassed open brake that was built at Eastleigh in 1933. It remained in service until preserved in 1965 by the Kent & East Sussex Railway. Following a major restoration on the railway, it was put back into service in June 2003 and was seen between duties, soaking up the summer sunshine beside the shed at Tenterden, on 28 July 2018.

The SR also initiated a pre-Second World War electrification programme. No. 11187 was one of those vehicles. It was built in 1938 at Eastleigh as a bogie driving motor brake third-class saloon that was part of a 750v third rail electric multiple unit. It was preserved in 1972 and placed in the owners' garden, in Cheshunt, until 1981. It currently resides on the East Kent Light Railway, at Shepherdswell, where it was pictured on 19 April 2014.

Built in 1946 as a 750v third rail electric multiple unit bogie trailer third-class vehicle, this was No. 10400. It was withdrawn in 1960, when it was converted to a de-icing trailer. In 2009 it moved to the former Finmere station site, at Newton Purcell in Buckinghamshire, where this picture was taken on 15 August 2010.

This carriage, No. 5768, was built as a composite at Eastleigh in 1947 and was purchased in 1968. It was stored in the open until 1971, when it was moved in poor condition to the Bluebell Railway. It was finally put into service on the railway in 1977. However, due to substantial rot having been found throughout the bodywork, it was taken out of service and, by 2010, it had begun a major restoration. It was pictured nearing the end of that process at Horsted Keynes on 9 October 2016.

Chapter 3

Irish Railways

The railways of Ireland were formed of eight major companies and forty-four minor ones. Whilst on the mainland the railways were grouped together under four companies in January 1923, this was not the case in Ireland – railways in Northern Ireland fell under the Ulster Transport Authority in 1948, whilst in 1924 it was deemed that all the companies in the Irish Free State (Southern Ireland) would be grouped together in 1925 under the banner of Great Southern Railways. One of the major companies, the County Donegal Railways Joint Committee, remained independent until closure in 1960. There were also twenty-five minor companies that, due to being regarded as light railways/ tramways, were never grouped into either the UTA or the GSR.

Opened in 1834, the Dublin & Kingstown Railway was the first railway to ever operate in Ireland. This vehicle is the only surviving piece of rolling stock from that railway, built in Dublin in 1838, and is a semi-open third-class carriage numbered 48. It was originally built to standard (4 foot 8.5 in.) gauge; however, in 1854 the lines of the railway were leased to the Dublin, Wicklow & Wexford Railway, who re-gauged everything to the now recognised Irish standard gauge of 5 ft 3 inches. It can be seen on display at the Ulster Folk and Transport Museum in Cultra, Northern Ireland, where it was pictured on 4 April 2015.

The Midland Great Western Railway was the third largest standard Irish gauge railway. Several carriages survive. This example, seen at the Ulster Folk and Transport Museum on 4 April 2015, is the best-known of those. Built by Joshua Dawson at Phibsborough works, just outside of Dublin, in 1844 as the private saloon/office of William Dargan (builder of the Dublin & Kingstown Railway), it was allocated No. 47. However, it is usually just referred to as 'The Dargan Saloon'.

The 3-foot gauge West Donegal Railway was formed in 1860. This vehicle was built in 1882 by the Birmingham RC&W Co. as a four-compartment tri-composite and was given No. 1. It was then rebuilt in 1910 as a first-class saloon with lavatory (also used as a directors' saloon). It was pictured at the Cultra Museum on 4 April 2015.

4 April 2015 again and at the Ulster Folk and Transport Museum is this 1887-built composite, No. 5. It was built by the Metropolitan C&W Co. for the 3-foot gauge Cavan & Leitrim Light Railway, which operated from 1887 until 1959.

Operating between 1891 and 1921 was the 3-foot gauge Tralee & Dingle Light Railway & Tramway. Seen at the Cavan & Leitrim Light Railway & Museum on 26 April 2018, was this Bristol C&WW Co. 1890-built, thirty-four-seat, third-class carriage that was given No. 6. In 1939 it was transferred to the West Clare Railway. In 1951 it was taken to Inchicore Works, where the body had rotted away. It was therefore replaced by one that was made from two scrapped Leyland Lion road buses, and then returned to the West Clare until that closed in 1961.

The M&GWR was formed in 1845. This five-compartment second-class carriage was built at the Ashbury Railway Carriage & Iron Company in Dublin in 1892 as No. 53. It entered preservation in 1985 and was seen inside the purpose-built museum at the Downpatrick & County Down Railway Preservation Society on 4 April 2015.

This 1892-built M&GWR five-compartment third-class carriage, No. 62, is unique by the fact that one of the compartments is a ladies only compartment. It was pictured on 5 April 2015 at the Railway Preservation Society of Ireland's headquarters at Whitehead, where it was undergoing a gradual restoration process.

Operating for fifty-four years, the County Donegal Railways Joint Committee ran an extensive 121 miles of 3-foot main line railways. It came into being in 1906 and lasted until 1960. Pictured on 6 April 2015, at the Foyle Valley Railway Museum in Londonderry, is this 1893-built County Donegal Railway five-compartment composite, No. 14.

Built by Ashbury in 1894 for the M&GWR, No. 25 is a five-compartment second-class carriage. It can currently be seen in the museum of the Downpatrick & County Down Railway Preservation Society, where it was pictured on 4 April 2015.

Another CDR vehicle found on 6 April 2015 at the Foyle Valley Railway Museum was this 1901-built Oldbury C&W Co.-built family saloon, No. 30.

Undergoing major restoration at the RPSI base at Whitehead, Northern Ireland, on 5 April 2015 was this 1906-built clerestory roof side corridor tri-composite brake, No. 861. It was built for the Great Southern & Western Railway of Ireland for use on the 'Rosslare Express' boat train services and it was the first carriage acquired by the RPSI in 1972.

Among the historical railway artefacts at the Ulster Folk and Transport Museum at Cultra is this sole surviving 1909-built five-compartment first- and second-class composite from the Dundalk Newry & Greenore Railway, which was seen on 4 April 2015. The railway was closed in December 1951 and was dissolved by mutual consent in 1957.

Built in 1911, at the Dundalk Works of the Great Northern Railway of Ireland, was directors' saloon No. 50. In 1953 it was used as the Queen's saloon in the Irish Royal Train; it was withdrawn in 1970 with preservation following three years later. It was pictured at the RPSI base in Whitehead on 5 April 2015.

Pictured on 4 April 2015, at the Downpatrick & County Down Railway in Downpatrick, was this Inchicore Works 1915-built GSWRI tri-composite, No. 2091. By the time it was transferred to the departmental fleet in 1967 it had been downgraded to a third-class vehicle and was used as a dormitory van, hence the steel panelling mid-way down the side.

Built by the GSWRI at Inchicore, in 1923, was this composite with an open saloon at one end, numbered 1097. During the war it was utilised as an ambulance coach and prior to its preservation in 1985, it was used as part of the Inchicore Works breakdown train. This photograph was taken at Downpatrick on 4 April 2015.

In July 1903 the Midland Railway acquired the Belfast & Northern Counties Railway (formerly the Belfast & Ballymena Railway), and the Northern Counties Committee came into being. Built in 1924, at Belfast York Road, was this side corridor open third-class carriage, No. 243. It was noted on 5 April 2015 at the RPSI base in Whitehead, where it was undergoing a slow rebuild.

Between 1888 and 1932 steam-powered trams operated between Terenure in Dublin and Blessington in County Wicklow. One of the later vehicles used was this 35 hp petrol railcar, No. 3, built by the Drewry Car Co. Ltd, London, in 1926. Upon the closure of the tramway this railcar was sent to a scrap merchant, where it was spotted and subsequently purchased by the CDR and re-gauged to 3-foot gauge. Its engine was removed in 1944 and it was used as a trailer car. It was pictured at Cultra on 4 April 2015.

It's not difficult to spot the origins of the vehicle in this 4 April 2015 Ulster Folk and Transport Museum photograph. It was built by the GNRI in 1926 at their Dundalk Works, as a half cab road bus and given a letter instead of a number ('E'). In 1936 it was fitted with rail wheels and became Railbus E2, before being loaned to the Sligo, Leitrim & Northern Counties Railway in 1938. It was rebodied and re-engined in 1947 and between 1958 and 1963 it was used as an inspection vehicle. It was withdrawn in 1965 before preservation in 1969.

It's a matter of a first and a last where this vehicle is concerned. Built as an articulated diesel railcar by Walker Brothers Ltd of Wigan in 1932, No. 1 had the honour of being the first such vehicle in Ireland. It was built for the 3-foot gauge Clogher Valley Railway. It also worked the last train on the CVR on 31 December 1941 (the railway closed the next day). It was pictured at the Cultra Museum on 4 April 2015.

This carriage saw service for forty-four years prior to being preserved in 1975. It was built by the NCC in 1934, as a bogie corridor brake third-class carriage, for use on the 'North Atlantic Express' services at Belfast York Road and was numbered 91. It was pictured on display on 15 June 1996 at the public open day celebrating the 150th anniversary of Inchicore Works, Dublin.

Built in 1934 as a collaboration between Walker Brothers of Wigan and the CDR at Dundalk Works, this was diesel railcar No. 12. The CDR were pioneers of diesel railcars in Ireland. This example was pictured at the Foyle Valley Railway Museum in Londonderry, on 6 April 2015.

The Great Southern Railways Company was formed in 1925. This bogie corridor third-class carriage, No. 1335, was built by the GSR at Inchicore Works in 1937. It was pictured on 15 June 1996 at the place it was built during an open day.

Pictured at Inchicore Works, during a public open day on 15 June 1996, this was GNRI 1938 first-class dining car No. 88. It was built at Dundalk Works for use on the Dublin to Belfast express services.

Another preserved GNRI carriage is this 1940-built third-class open brake, No. 114. In 1958 it was converted for railcar working and had a vertical steam boiler fitted to one end. On 5 April 2015, it was awaiting restoration at the RPSI base in Whitehead.

No. 18 is another CDR/Walker Brothers diesel railcar that has been preserved. It was built in 1940 at Dundalk and was rebuilt ten years later, following a fire. It can be seen operating during the summer months at Fintown on the An Mhuc Dhubh Railway, where it was pictured during the closed season inside the shed on 29 April 2011.

Chapter 4

Isle of Man Railway

The Isle of Man Railway Company was formed in 1870 and operated between 1873 and 1977, when it became nationalised as the Isle of Man Railways, and later Isle of Man Passenger Transport. The locomotives and stock became the property of the 3-foot gauge Isle of Mann (note spelling) Victorian Steam Railway Company Ltd in 1965 which, in 1978, became the island's government-owned Isle of Man Steam Railway, which operates as a tourist railway.

Built by Brown, Marshalls & Co. Ltd in Saltley, Birmingham, in 1881, this was six-compartment third-class carriage No. F10. It was seen inside the shed at the Douglas depot. All photographs in this chapter were taken on the same date of 19 September 2006.

Brown, Marshalls only built a small number of carriages used by the railway and this 1894 brake third-class carriage, No. F18, is another one of those. It is pictured in the Railway Station Museum at Port Erin on the island.

F25 was built by another Birmingham company, namely the Metropolitan Carriage & Wagon Co. It was built in 1896 as a brake third-class vehicle, which later became a third-class. It was seen at the Port Erin Railway Museum.

The MC&W Co. built this third-class saloon, No. F31, in 1905 and it was noted inside Douglas shed in between passenger services.

Built in 1905 by the MC&W Co., as No. F36, this is a composite saloon with a handbrake. It was part of the island's Royal Train and was used by the Queen Mother during 1963. It is used nowadays for carrying dignitaries and it is kept in the Port Erin Museum, where it was pictured.

Inside Douglas shed undergoing a rebuild is this 1908 MC&W Co.-built bogie composite brake, No. F43.

This composite carriage, No. F62, was built in 1926 by MC&W Co. when two four-wheeled carriage bodies (of unknown origin) were joined and mounted onto a new underframe. In need of some TLC, it is seen inside the Museum at Port Erin.

Seen on display inside the Port Erin Museum is this MC&W Co. 1926-built governor's saloon (No. F75), which was used by the Duke of Sutherland for the railway's opening. Like the previous picture, it was built by joining two four-wheeled carriages bodies, of 1893 origin, and mounting them onto a new underframe.

Built by Walker Brothers of Wigan in 1947, for the County Donegal Railways, this was diesel railcar No. 19. When the CDR closed in 1960 the Isle of Man Steam Railway purchased two of these vehicles. It was seen undergoing a rebuild and restoration inside the Douglas workshops.

CHAPTER 5

Pullman Carriages

The history of the Pullman Car Company was covered extensively in *Railway & Tramway Bodies – Another Life*. Suffice to say that it is an internationally recognised name when it comes to luxury high-class train travel.

No. 47 *Alicante* was built in 1912 as a kitchen car by Cravens of Sheffield for the Pullman services on the SECR. In 1962 it was transferred along with five others to Marazion, near Penzance, where it was pictured on 24 July 1995, for use as camping coaches by the Western Region Staff Association.

Another SECR Pullman kitchen car, noted at Marazion on 24 July 1995, was No. 50 *Mimosa*, which was built by the BRCW in 1914. This example, the previous one and one other were taken to the former Petworth station, in West Sussex, in November 1998, where they are used as holiday accommodation.

On display inside the NRM at York, on 26 October 2013, is this 1914 BRCW-built SECR Pullman services parlour car, No. 59 *Topaz*. It was withdrawn for preservation in 1969.

In 1921 Claytons of Manchester built 114 *Car No. 14* as a third-class kitchen car for the LBSCR Pullman services. It was built utilising a redundant 1918 LNWR ambulance coach underframe. In 1947 it was rebuilt as an observation saloon and, in 1969, it went to America as part of the 'Flying Scotsman' tour. It was purchased in 1973 and was used as a restaurant/boardroom in San Francisco, until it was repatriated and restored in 2007. On 31 March 2012, it was seen at Norden on the Swanage Railway at the rear of a passenger train.

Another LNWR ambulance coach underframe, of 1914, was used by Claytons in 1921 to build kitchen car 135 *Elmira* for the SECR Pullman services. It was preserved in 1968 and was seen on 19 July 2015 at Eskdale, on the Ravenglass & Eskdale Railway in Cumbria, where it is used as holiday accommodation.

The BRCW built this gangwayed full brake, No. 154 *Flora,* for the SR in 1923. After languishing at Marazion (where it was pictured on 24 July 1995) for several years, it was taken, along with the first two vehicles in this chapter, to Petworth station to be used as holiday accommodation.

Another carriage BRCW built for the SR in 1923 was gangwayed full brake No. 156 *Montana*. It was withdrawn in 1969 and taken to Cambridgeshire, where it was used as a dwelling. By June 2007 it had arrived at Petworth station, where it was pictured, as holiday accommodation on 9 July 2011.

This 1923 Clayton-built third-class kitchen car 157, *Car No. 54* (later named *Fiona*), was built for the LNER. It was preserved in 1963 and, on 9 October 2016, it was seen undergoing a major restoration rebuild at Horsted Keynes on the Bluebell Railway.

Built in 1932 by the MC&W Co. for the SR was No. 179 *Hazel*, which was a kitchen car. In 1972 it was purchased by the owners of the Black Bull Inn at Moulton, to be used as a restaurant – it was pictured there on 9 July 2003.

No. 282 *Doris* was built for the SR in 1932 by the MC&W Co. as a kitchen parlour trailer car. In 1972 it was purchased for preservation by CIL Shopfitters Ltd, taken to their Finsbury Park premises, put onto a siding beside the mainline and used as a boardroom. It was pictured there on 12 April 1996.

This 1932-built SR EMU Driving Motor Brake Parlour Car 289, *Car No. 89,* was bought in 1972 and taken to the Little Mill Inn at Rowarth, in the north of Derbyshire. In 1976 it was auctioned and was due to move to Windsor Safari Park, but removal was too costly. It therefore remained at Rowarth as bed & breakfast accommodation and was pictured there on 18 July 2012.

Chapter 6

Miscellaneous Carriages

Aside from the carriages from the Big Four and pre-grouping companies and others mentioned, there are numerous overseas (usually narrow gauge) carriages that were built prior to 1948 that are preserved across the UK.

Plinthed on a short section of track at the Chatham Historic Dockyard, on 20 April 2014, was this War Department MC&W Co.-built 1885 clerestory-roofed saloon (WD Registered No. RCP003020). It was built for the Sudanese Suakin to Berber Railway and was used by Lord Kitchener. It was repatriated in 1898 and used by the military until preservation in 1990.

Coaches built by the Metropolitan Railway are easily distinguished by the circular tops on the doors. Pictured at the Covent Garden London Transport Museum, looking resplendent on 15 December 2012, was No. 400, which was a second-class vehicle built at the railway's workshops in Neasden during 1899.

No. 21, a two-compartment second-class balcony carriage, was built in 1900 by Grazer W&M for use on the 2-foot 6 in. gauge Zillertal/Zillertalbahn in Austria. In 1968 it was donated to the Welshpool & Llanfair Railway, where it was pictured on 7 September 2008 bearing its new number of B14. It was rebuilt in the early 1990s as it was becoming unsuitable for use on passenger trains, due to deterioration of the bodywork.

B27 was another Grazer-built vehicle, this time for the Austrian OBB Gmünd to Gross Gerungs Railway (which was also 2-foot 6 in. gauge). It was built as a two-compartment balcony carriage in 1906 and numbered 3660. The railway was taken over by the Zillertal in 1969 and in 1975 they donated this vehicle to the Welshpool & Llanfair Light Railway, where it was pictured on 7 September 2008. Like the previous picture, this vehicle was also rebuilt in the early 1990s.

This second-class open balcony carriage was built in 1926 for the NSB (Norwegian State Railways) and was numbered 18803. During the 1990s it was at the Colne Valley Railway. It was moved from there in 2000 to its current location at the former Hole station in Devon, where it was pictured, on 24 May 2015, being used as an office/residential accommodation.

This carriage was built by the MC&W Co. in 1927 for the Rhodesia Railways and numbered 1808. It was used as an immigrant sleeping car and transported people to Rhodesia. Prior to its withdrawal it operated on the Zambezi Sawmills Railway and was preserved in 1973. It was seen at the East Somerset Railway, in Cranmore, on 15 July 1995.

No. 412 was built by Skabo, in Oslo, for the NSB, in Norway, in 1928, as a clerestory-roofed five-compartment corridor brake. It was seen at Blunsdon on the Swindon & Cricklade Railway on 6 March 2010, where it is used as a buffet/cafeteria.

With its windows boarded up to prevent vandalism, this 1930-built un-numbered saloon third-class carriage was seen at Blennerville, the site of the ill-fated Tralee & Dingle Light Railway in Co. Kerry, Ireland, on 15 April 2017. This was one of four coaches that were purchased from Spain and re-gauged to 3-foot gauge to run on the railway.

This vehicle was built in 1931 for the SNCB (Belgian State Railways) as No. 42170, a second-class open. It was noted on 16 February 2013 at Wansford, on the Nene Valley Railway, having been preserved in 1990. It is the only one of its type in the UK.

1934-built SNCB first-class open No. 21013 is seen here at Wansford, on the Nene Valley Railway, on 16 February 2013, following its preservation in 1990. The interior at one end as been removed for use as a lounge and entertainment area.

This is another preserved SNCB 1934-built vehicle, this one being a composite and numbered 23009. Wansford is the location, 16 February 2013 the date. It is believed that this vehicle is the only surviving one of this type.

This carriage was built at the Remarul locomotive works, Romania, in 1935, as a post wagon for the CFR (Romanian State Railways). It was rebuilt as a disabled-access passenger vehicle. It can be seen on the South Tynedale Railway, in Cumbria, and was pictured at Alston on 18 July 2015.

This second-class open, No. 22311, was built in 1938 for the Finnish State Railways. It was pictured on 23 June 2012 at the Snibston Discovery Park in Colville, Leicestershire. Upon closure of this site in 2017, it was moved to Tetbury, in Gloucestershire, where it is used as a café and workshop.

The Compagnie Internationale des Wagons-Lits (International Sleeping Car Company) was founded in Belgium in 1872 and quickly established itself as a premier provider and operator of luxury sleeping and dining cars. No. 2757 was built in 1939 as a restaurant car and was noted on 14 February 1996 at the GWR Society's Southall base.

Chapter 7

Rescued, Restored & Running Again

There are many vintage carriage bodies dotted across the country and a good number of those have been rescued and restored back to their former glories as passenger carrying vehicles, with the bodies mounted and fixed onto either new-build or donor (often shortened) underframes. The heritage preservation railways and museums are finding that they are being bequeathed even more of these bodies by their owners and therefore the demand for suitable underframes is at a premium.

Built in 1844 for the Manchester & Birmingham Railway as No. 1, this was a three-compartment first-class carriage. It was transferred to the Shrewsbury & Hereford Railway, where, in 1874, it derailed and damaged one end; the body was grounded until rescued in 1986. It is seen at the Manchester Museum of Science and Industry on 22 February 2014, where it has been mounted onto a four-wheel wagon underframe and restored.

Built at Oldbury in 1864 for the Isle of Wight Railway, No. 10 is a three-compartment composite. Following withdrawal, its body was grounded in 1923 and was rescued by the island's steam railway in 1983, where it lay in a field for another thirty-five years. It was finally restored in 2018, with a new wooden underframe constructed, and featured in the Channel 4 Series *Great Rail Restorations*. It was pictured on 29 June 2019 in the railway's shed.

No. 949 was built by the MC&W Co. in 1881, for the LBSCR, as a three-compartment brake third-class carriage. Withdrawn in the early 1920s, the body was used on a farm as a lambing shed until 1998, when it was donated to the Bluebell Railway. Seen at Horsted Keynes on 9 October 2016, it was near to the completion of its restoration.

Sold to the Kelvedon & Tollesbury Railway in 1928, this was 1884-built GER Wisbech & Upwell tramcar No. 7. It was withdrawn in 1951, and the body was used as an onion store until 2002. It has been restored with a brand-new underframe and bogies built in 2004 and was seen on 4 April 2010 at Weybourne on the North Norfolk Railway. This vehicle is the twin to the one featured in the Ealing comedy *The Titfield Thunderbolt*.

With its body being used as a cricket pavilion for fifty-six years until preserved in 1986, four-compartment first-class No. 4, built at Highbridge, in 1886, for the Somerset & Dorset Joint Railway, was noted at Norton Fitzwarren on the West Somerset Railway during their annual steam fayre, on 1 August 2015, following its restoration.

Withdrawn in 1932, the body of this 1886 Derby-built Midland Railway picnic saloon, No. 1616, has been used for many different purposes: a workshop/mess room for the signal and telegraph department, a clubhouse, a garden shed, a wedding reception venue and a TV and radio repair workshop. Noted at Weybourne on 4 April 2010, it was mounted onto a donor underframe from a parcels van whilst undergoing a complete restoration.

Second-class brake carriage No. 108 was built at the LCDR Longhedge workshops in 1888. Following withdrawal in 1921, the body was used as a bungalow until preserved in 1986 and restored. It was noted inside Rolvenden carriage shed on the K&ESR, on 28 July 2018, in between duties.

Surprisingly, No. 127, which is an 1889 North Staffordshire Railway-built four-compartment first-class carriage, has spent longer in preservation than it did in service. It was withdrawn in 1892 and its body was used as a store shed on a farm for eighty-eight years until acquired. It was noted at the Foxfield Railway, at Blythe Bridge, on 12 September 2015, fully restored to its former original glory.

No. 159 was built by the LNWR, at Wolverton, in 1890, as a clerestory-roofed corridor restaurant car. It served as an ambulance car in Earl Haig's First World War train, and was withdrawn in 1921; the body was sold as a holiday home in Bognor Regis. It was rescued from there in 1978 and restored. On 26 July 2003, it was pictured during an open day at Doncaster Works.

Cravens of Sheffield built this four-compartment first-class carriage, No. 353, in 1892 for the Metropolitan Railway. In 1907, it was transferred to the Weston, Clevedon & Portishead Railway, where it remained until closure in 1940. The body was sold and used as a garden shed for another thirty-four years until it was preserved. On 28 July 2018, it was seen at Rolvenden carriage shed on the Kent & East Sussex Railway.

No. 233 is an 1897 Swindon-built GWR clerestory-roofed royal saloon, built for Queen Victoria's Diamond Jubilee. For fifty-two years, following withdrawal, it was used as a holiday home on the cliffs at Aperporth, until it was preserved in 1982. On 2 June 2012, it was noted on display, following restoration at the GWR Museum, inside the former works building where it was built 115 years before.

In 1899, Cravens of Sheffield built family saloon No. 1509, which was used on the MS&LR until 1940, when it was withdrawn and used by the Home Guard. After the hostilities had ended, the body was used as a holiday home in Filey. In 1989, it was acquired by the Tanfield RPS and mounted onto a parcels van underframe. It was noted at Marley Hill on 8 July 2003.

Blackpool club car No. 47 was built for the L&YR at Newton Heath, in 1912, specifically for carrying businessmen between Blackpool and Manchester. It was withdrawn in 1951 and its body was used as a cricket pavilion in Derbyshire. It was purchased for £1 by the Keighley & Worth Valley Railway in 1993, placed onto a donor underframe and restored. It was seen at Oxenhope, on 26 October 2013, and is the only surviving L&YR club car.

Pictured at Taunton station whilst working a VSOE Pullman Services railtour on 27 April 2013 was 1932-built SR Pullman kitchen car No. 284 *Vera*. It was withdrawn in 1972 and promptly acquired by a private owner who used the body as a lodge for shooting parties in Suffolk. It was purchased by VSOE in 1985, who fitted new bogies and restored it back to its former glory.

Chapter 8

Preserved, But Now Gone

As has been pointed out in this book, some preservation railways/museums don't always have the resources or money to restore some of their stock and it is left out in the open to the elements and deteriorates. When this happens and the vehicle is deemed too far gone to save, then all reusable parts are removed and used to keep other carriages going and the original vehicle is scrapped. Although sometimes another reason for a vehicle being scrapped can be bought about when the vehicle is victim of arsonists or repetitive vandalism.

In 1962, six Pullman carriages moved to Marazion, near Penzance, to be used as camping coaches. In chapter five I detailed three of those vehicles that had moved to Petworth station; the first pictures in this chapter (all taken on 24 July 1995) detail the other three. The years of open Exposure to the salt air from the Cornish coast took its toll on these vehicles and, having been robbed of all reusable parts for the other three vehicles in August 2003, the local authority instructed that their remnants be removed and were therefore scrapped. No. 97 *Calais* was built in 1921, by the BRCW, for use by the SECR on their Pullman services.

No. 153 *Aurora II* was built by the SR in 1923. It was heavily vandalised and salt damaged, before being set on fire by arsonists in February 2002. Its remains stayed in situ until removed 18 months later.

Juno II, the last of the three vehicles, was again built by the SR in 1923 and was numbered 155. Salt damage, constant vandalism and being robbed of all reusable parts sealed its fate and it was scrapped.

This 1925 Doncaster-built LNER first-class sleeping car was numbered 1237 and was used as a staff and dormitory coach until 2001, when it moved to Bere Ferrers Station Museum, where it was pictured on 21 April 2003, showing damage from an arson attack. It was finally broken up in early 2006. Note the fire-charred wagon to the left of it.

Within two years of this 21 March 1996 picture, taken at Dunster on the West Somerset Railway, this 1925-built LSWR Ironclad third-class corridor brake, No. 3205, had its body removed and scrapped. The underframe was sold to Railholidays UK Ltd and was placed under a GWR carriage body. It is being prepared in readiness for use as holiday accommodation at St Germans station in Cornwall.

Built by the GWR at Swindon, in 1927, as No. 4926, was this corridor brake third-class carriage. During the 1960s, it was converted for use as a tunnel inspection staff coach. It was pictured in Bristol East yard on 8 April 1999 and was donated to a heritage railway for preservation in 2003. However, before it could be moved it was the target for an arsonist, in May 2004, and was finally broken up on site by August of that year.

Corridor composite No. 9389 was built by the LMSR at Wolverton in 1930. It was used as a staff coach until 1993, when it was acquired for preservation by the Ystwyth Valley Railway and taken to their base at Lanilar in 1997. The railway never took off and, on 21 April 2003, I pictured the carriage there looking sad and forlorn with all its windows smashed. It was broken up on site by December 2007, leaving just a pile of debris there by September the following year.

No. 9605 was built at Swindon by the GWR as a restaurant composite, but you'd never guess that from looking at this 25 August 1997 photograph that I took at Big Pit sidings on the Pontypool & Blaeanavon Railway. Due to no work being carried out on it in 2009, it was offered for sale, but, with no takers, it was scrapped on site in 2013.

Built by Metropolitan-Cammell in 1931, this was 110V overhead electric Manchester, South Junction & Altrincham Railway trailer vehicle No. 114. It was one of three MSJ&AR carriages preserved in 1971 by Derby Museum, prior to them being moved to the Midland Railway Centre at Butterley (where it was pictured on 31 August 1998). This one was the worse of the three and was offered for sale, in 1999, before it was scrapped on site during 2006/07.

DS 70244 was the departmental number of this 1935-built LMSR corridor brake third-class carriage, whilst it was being used as a staff mess and tool coach; No. 6389 was the original number when it left Wolverton Works. Pictured on 26 February 1996, at the Avon Valley Railway headquarters in Bitton, where, during 2007, it was the victim of an arson attack and the body was destroyed. The underframe was finally cut up in 2011.

No. 22324, a second-class open, was built in Helsinki, in 1935, for the Finnish State Railway. It was preserved at the Epping & Ongar Railway, where it was pictured on 9 August 2007. It was externally cladded to represent the state carriage in the 2012 Keira Knightley film *Anna Karenina*. It was deemed surplus to requirements and, along with another Finnish carriage, it was scrapped in 2013.